A revised Key to the adults of the British Species of Ephemeroptera

WITH NOTES ON THEIR ECOLOGY

by

D. E. KIMMINS

*Formerly of the Department of Entomology,
British Museum (Natural History)*

FRESHWATER BIOLOGICAL ASSOCIATION
SCIENTIFIC PUBLICATION No. 15
Second revised edition 1972

FOREWORD

For the most part this Key is a reprint of Mr Kimmins' earlier Key (No. 7 in the Association's series of *Scientific Publications*) the stock of which is exhausted. The necessity to reprint has, however, provided the opportunity for a number of additions and corrections including the addition of a species recently added to the British list, and the arrangement of the text and figures has been improved.

The keys to the families and genera of the nymphs, which were included in the earlier edition, have been omitted here. This is because a complete key to the species of the nymphs is in preparation as a forthcoming member of the series. Had it not been for the exhaustion of the stock of No. 7 and the continued steady demand for it, the production of this new edition would have been deferred until the key to the nymphs should be ready, but in the circumstances it was thought desirable to re-issue this Key to the adults without delay.

The Association is indebted to Mr Kimmins for his continued interest and for the time he has given to the preparation of this revised edition. It has unfortunately been necessary to raise the price.

THE FERRY HOUSE,　　　　　　　　　　　　　　　　　H. C. GILSON,
February, 1954.　　　　　　　　　　　　　　　　　　　　*Director.*

This, the second edition of Mr Kimmins' revised Key, differs little from its predecessor. Another species, *Baetis digitatus* Bengtsson, has been added tentatively to the list; more material, especially of adults, would be very welcome. The nomenclature has been brought up to date, and the families have been rearranged in the order adopted by Edmunds, Allen & Peters (1963); this has involved the renumbering of the figures. At the same time the notes on ecology and distribution have been brought into the body of the Key.

THE FERRY HOUSE,　　　　　　　　　　　　　　　　　H. C. GILSON,
February, 1972.　　　　　　　　　　　　　　　　　　　　*Director.*

SBN 900386 17 7

FOREWORD TO THE FIRST EDITION

This pamphlet should appeal to fly-fishermen and those interested in fishery development as much as it will to entomologists. It provides for the first time a means of identifying reliably all the species of Ephemeroptera known in Britain, without the necessity of referring to a multitude of technical papers. In relation to freshwater fisheries this is perhaps the most important order of aquatic insects. Although only forty-seven species are yet known to live in Britain, most of them occur in enormous numbers where conditions are suitable and in all stages of their life history they provide food for fish. Thus a better understanding of the habits of life of these insects should help in producing better fish as well as in adding to the interest and efficiency of angling.

There is no satisfactory English name for the group as a whole. The term "Mayflies," though sometimes used by entomologists for the group, is restricted to the species of *Ephemera* by anglers, who use the terms "dun" and "spinner" for the subimago and imago respectively of all species. To avoid confusion therefore, the group is referred to throughout the pamphlet by the scientific name.

Mr Kimmins, the British Museum specialist on Ephemeroptera, is the person best qualified to prepare this publication, and we are greatly indebted to him for having devoted much time to it. While working at Wray Castle since the spring of 1941, he has lost no opportunity of becoming familiar with the aquatic insects of the Lake District in a state of nature as well as pinned in cabinet drawers. Moreover he has made full use of extensive collections and notes on the group made by Dr T. T. Macan, the Freshwater Biological Association's entomologist now on active service.

It is hoped that this pamphlet will stimulate scientific work and observation on the Ephemeroptera. Even the list of British species is probably incomplete, as witnessed by the addition to it of three species during the past year or so, one by Macan from the Hampshire Avon, and two by Kimmins from the Lake District. The life histories of but few species are completely known, and much research is required before the nymphs of all species can be identified without fail. Therefore, though complete for the time it is written, this publication may have to be revised and extended in the future.

The Association is much indebted to the Royal Society for grants toward the cost of publishing this and other numbers in the series of *Scientific Publications*.

WRAY CASTLE.
August, 1942.

E. B. WORTHINGTON,
Director.

CONTENTS

	PAGE
INTRODUCTION	5
CHECK-LIST	12
COLLECTING AND PRESERVING	14
FISHERMEN'S NAMES	17
KEY TO FAMILIES	21
KEYS TO GENERA AND SPECIES, WITH NOTES ON ECOLOGY	25
REFERENCES	73
INDEX TO FAMILIES AND GENERA IN THE KEYS	75

INTRODUCTION

This group of insects originally formed a section of the old Linnaean Neuroptera but is now generally regarded as a distinct order. Although in the past this order has been known under various names such as Plectoptera, Ephemerida and Agnatha, it is now usually termed Ephemeroptera. In recent classifications the Ephemeroptera are placed near the Odonata (Dragonflies), being considered more specialised than the Plecoptera (Stoneflies). In number of species the order is but a small one, only about nine hundred having been described. Geological evidence indicates that it was once proportionately much more abundant in species and that it has steadily lost ground since the time of the earliest recognisable fossils. These early forms, which are found in the Permian shales of Kansas and Archangel, differed especially from modern Ephemeroptera in having both pairs of wings of about equal size, and all three pairs of legs long and slender. Reduction of the size of the hind wing is to be found in certain Jurassic fossil Ephemeroptera. Adult specimens of modern genera have been recognised amongst the Baltic Amber fossils of Tertiary times.

Ephemeroptera have retained some very primitive characters, notably in the sexual organs of both sexes, in which the ducts are paired throughout their length and terminate in separate openings. On the other hand, the possession of a subimaginal stage, at the end of which the winged insect undergoes a further ecdysis before attaining sexual maturity, is a specialised feature unique in insects.

In the British Isles the Ephemeroptera have probably attracted less attention from entomologists than from fly-fishermen, and in consequence there has been no comprehensive systematic work including all the British species since Eaton's big monograph of the species of the world (1883-1888) and his brief synopsis of the British species, also in 1888. Various handbooks, with keys to the European (particularly German) species, have been produced on the Continent, notably by Klapálek in the Süsswasserfauna Deutschlands (1909), by Ulmer in the Tierwelt Mitteleuropas (1929) and Schoenemund in the Tierwelt Deutschlands (1930). These suffer (from the point of view of the British collector) from the joint disadvantages of including species not of the British fauna, and of omitting certain British species which are not recorded from Central Europe. The present keys and figures will, it is hoped, make good these deficiencies, and render it easier for the British worker to identify his captures. In certain genera, particularly *Baetis*, it has not been found possible to construct keys which will separate the females satisfactorily, and even the separation of the males of this genus is not as definite as could be wished. In *Baetis*, the male genital structures are regrettably uniform in pattern, and the venation of the hind wing, which has been much used as a specific character in this genus, has proved to be unstable and unreliable in the finer shades of difference. Further breeding work on this genus may assist in clarifying the situation.

All the figures (unless otherwise stated in the legend) are original and drawn by the author from British samples.

Table 1. *Nomenclature of wing veins.*

Needham	Ulmer, Schoenemund	Eaton
C	C	1
Sc	Sc	2
R_1	R_1	3
R_2	Rs	4
R_3		5
R_4	M_1	6
R_5	M_2	
M_1	Cu_1	7
M_2	Cu_2	
Cu_1	A_1	8
Cu_2	A_2	9
A_1	A_3	

STRUCTURE. In the classification and identification of the Ephemeroptera, the more important characters are to be found in the wings, in the male accessory genitalia and to a lesser degree in the proportions of the tarsal segments. The colour of the eyes in living males, the colour pattern of the abdomen and of the subimaginal wings also offer useful characters for the separation of species. The margins of the wings are fringed with hairs in subimagines (and in the imagines of the Caenidae).

The wings and their venation, being of primary importance in the classification, will be first considered. In general, there are two pairs of wings, usually with numerous cross-veins. The hind wing is always much smaller than the fore wing, and in certain genera is either minute or absent. The nomenclature of the various veins has been the subject of much discussion, and various interpretations and systems of notation have been put forward. In the present work, the system proposed by Needham in "The Biology of Mayflies" has been adopted and Table 1 has been drawn up to show the equivalent designations used by Ulmer and Schoenemund, and also the numerical notation of Eaton's monograph.

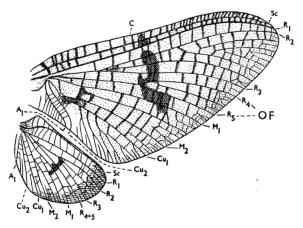

Fig. 1. The wings of *Ephemera danica* × 4·5. C, costa; Sc, subcosta; R, radius; OF, outer fork; M, media; Cu, cubitus; A, anal veins.

The *costa* (C, fig. 1) forms the front edge of the wing, and behind it runs the *subcosta* (Sc), more or less parallel to it and to the anterior branch of the *radius* (R_1). The posterior branch of the *radius*, which in the fore wing is generally more or less detached from the anterior branch, forks very near its origin, each fork again dividing before the wing margin to form the veins R_2, R_3, R_4, and R_5. In the areas between the main veins in the outer part of the wing, particularly between R_2 and R_3, are a number of additional veins (*intercalary veins*). The area between R_4 and R_5 is termed by Needham the *outer fork* (OF). The *media*, or median vein (M) has only one fork, its two branches being M_1 and M_2. The *cubitus* is also bifurcate, the anterior branch (Cu_1) being usually longer than the posterior branch (Cu_2). *Anal* veins (A), of which there are one or more, are usually weak and irregular. The spaces between the veins, and also the cross-veins in general take the name of the vein in front of them; thus the area between the costa and subcosta is termed the costal area and the cross-veins in it the costal cross-veins. The pterostigma is a portion of the costal and subcostal areas towards the apex of the wing; it is sometimes more heavily pigmented, and the cross-veins in it may be thickened, or more numerous, or both. The strong cross-vein at the base of the costal and subcostal areas is variously termed the basal, humeral, or great cross-vein, or costal brace. In the hind wing the costa and subcosta are often more or less fused except near the base.

The male accessory genital structures (fig. 28) consist of a pair of segmented forceps, or "claspers," arising from a plate, or "forceps-base," pertaining to the ninth sternite. The number of segments composing the forceps is variable, usually three or four, but the segmentation is often incomplete, especially between the basal and second segments, the latter being the longest and strongest. Above the forceps-base arise the paired penis-lobes, which differ considerably in form in the various families. Penis-lobes sometimes become distended or distorted at death, causing an atypical appearance. In some they are well sclerotised and offer good taxonomic characters; in others, Baetidae for example, they are reduced to slender, sinuous struts, largely internal, supporting the membranes of the penis.

In the female the paired oviducts open between the seventh and eighth sternites, the seventh sternite sometimes produced into a weak subgenital plate, the ninth sternite variously produced in a ventral plate.

The head has the mouthparts atrophied and functionless. Compound eyes of the male larger than those of the female, often differently formed, sometimes divided, the upper part elevated and turret-like and known as the turbinate eyes; three ocelli; antennae short, with two segments and a terminal bristle. Legs feeble, male front legs much longer than those of the female; tarsi with five, four or three free segments, one or two segments being fused to the tibia in the latter cases. The length and proportion of the tarsal segments are subject to abnormalities, particularly when a leg of the nymph has been lost and regenerated. Tarsal claws two, the members of a pair sometimes dissimilar in form. Abdomen with ten segments, tergites and sternites (except of the tenth segment) subequal in size; apex of the abdomen bearing three or two long, many-jointed tails or setae, the middle one being sometimes so much reduced in certain genera as to be to all intent absent.

LIFE HISTORY. The early stages of all Ephemeroptera are passed in fresh water. Eggs are laid in one of three ways:

(a) by the female washing off a few eggs at a time, dipping the tip of her abdomen at intervals while flying over the water, or by actually settling on the water for short periods, as in *Ephemera, Leptophlebia, Ecdyonurus,* etc.;

(b) by extruding the eggs in one cluster, and depositing them all at once, either by dropping them or by dipping and washing them off, as in *Ephemerella, Centroptilum*;

or (c) by the female crawling down projecting sticks or stones, through the surface film and depositing her eggs on under-water objects, as in certain species of *Baetis*. Unattached eggs sink to the bottom and adhere to whatever they touch, and remain thus until they hatch, a period lasting from a week or two to as much as several months.

The nymphs are of diverse types and many forms show decided adaptation to the type of habitat in which they live. Needham has divided them into two main groups, each sub-divided into three sections:

I. Still water (static) forms:
- (a) Climbers among vegetation, agile, streamlined forms — *Siphlonurus, Cloeon,* etc.
- (b) Sprawlers upon the bottom, silt-dwellers — *Caenis*.
- (c) Burrowers in the bottom — *Ephemera*.

II. Rapid water (lotic) forms:
- (d) Agile, free-ranging, streamlined forms — *Baetis, Ameletus*.
- (e) Close-clinging, limpet forms found under stones — *Rhithrogena, Ecdyonurus,* etc.
- (f) Stiff-legged, trash, moss and silt inhabiting forms — *Ephemerella*.

It is difficult, however, to draw a hard and fast line between still water and running water forms, as the controlling factor is not only the movement of the water, but its degree of aeration and possibly its range of temperature. Pools in rapid streams often approximate to still water conditions, and one must consider the micro-habitat rather than whether a species occurs in a stream or in still water. Certain species such as *Ephemera danica* and *Siphlonurus lacustris* occur in both still and slowly running water, and in the case of *S. lacustris* nymphs have also been found in small becks. Other species, such as *Heptagenia lateralis* and *Ecdyonurus dispar,* occur not only in rapid streams but also on the wave-washed stony shores of large lakes, where the wave action provides the necessary aeration.

The nymphs of Ephemeroptera feed mainly upon vegetable matter, filamentous algae, diatoms and fragments of higher plants. Some are believed to be partly carnivorous.

Some or all of the first seven abdominal segments carry lateral tracheal "gills," which differ considerably in form according to the genus. The lamellate type may be single or double and are sometimes accompanied by tufted gills at their bases. In some genera (not British) tufted gills are found at the bases of the maxillae and also at the base of each fore leg. At one time it was believed that all these gills had a respiratory function. Experimental work (Munro Fox and others) has shown that in those genera with plate-like gills, provided that a flow of water be maintained, there is no difference in oxygen consumption between normal nymphs and those from which the gills have been artificially removed, and that oxygen must therefore be absorbed directly through the integument of the body. The function of the gills in these cases would appear to be rather that of maintaining a flow of water over the surface of the body.

As the nymph grows, it moults frequently; the wing rudiments first appear when it is about half-grown and increase in size at each moult. The full-grown nymph gives rise to a subimago or "dun." In this stage it is fully winged, but the wings are dull and opaque, with a fringe of minute hairs, and the legs and caudal setae are not of the full length. In those genera in which there are only two setae in the adult, the median seta degenerates at the final nymphal moult and is practically absent in the subimaginal stage.

Transformation to the subimago may take place on the surface of the water, the subimago emerging from the floating nymphal skin; the nymph may crawl partly or entirely out of the water on a stick or stone; or emergence may take place on a stone beneath the surface, the subimago either crawling out on a projecting stone or floating to the surface. The time spent in preparation for the operation of emergence appears to vary with the species and local conditions, but the actual emergence is usually completed in a short space of time, variously reported to be from a few seconds to several minutes. After emergence, the subimago flies away from the water and takes shelter amongst vegetation and there, after a period of rest, it once more sheds its skin and becomes an imago or "spinner," capable of reproducing its species. This final moult of the subimaginal Ephemeropteron, which is unique in insects, does not occur in some non-British species. Shortly before the transformation to the imago subimagines of some species depress the wings parallel to the surface upon which they are resting.

The males of many species gather in large swarms at certain times of the day to take part in their beautiful rising and falling mating flight. This swarming flight does not necessarily occur only in fine weather; I have seen the males of *Cloeon dipterum* dancing in fine drizzle. Mating takes place in the air, often high up. The male grasps the female from beneath, with his long fore legs curved back over her thorax, his abdomen turned up so that the forceps clasp around her abdomen, the pair meanwhile slowly losing height. Copulation lasts for a short period only and is usually completed before the ground is reached.

The life cycle of many Ephemeroptera takes one year, that of others (*Ephemera*) takes two years, and others accomplish two generations each year, specimens from the generation hatched in the autumn often being smaller than those in the spring. In yet other species there may be three generations in two years.

ACKNOWLEDGEMENTS. My thanks are due to Dr E. B. Worthington, in whose laboratory much of the work of preparing these keys was done, to Dr W. E. Frost, of the Freshwater Biological Association and to my colleague Dr W. E. China, all of whom read through the original manuscript and made many helpful suggestions. I should also like to thank Mr H. C. Gilson for the work he has done in re-arranging illustrations and preparing this revised edition for press, and Dr T. T. Macan for the use of his notes and collections of Lake District Ephemeroptera in the preparation of the first edition and for helpful suggestions for this revised edition.

CHECK LIST

In the following revised check-list of the British species the families have been arranged after Edmunds, Allen & Peters (1963). Kloet & Hincks in their recent list (1964) show forty-six species. One, *Caenis moesta* Bengtsson, was accidentally omitted, and *Ecdyonurus forcipula* (Pictet) is also not mentioned.

Family SIPHLONURIDAE
  SIPHLONURUS Eaton, 1868
    1. *armatus* Eaton, 1870
    2. *lacustris* Eaton, 1870
    3. *linnaeanus* (Eaton, 1871)
  AMELETUS Eaton, 1885
    4. *inopinatus* Eaton, 1887

Family BAETIDAE
  BAETIS Leach, 1815
    5. *fuscatus*[1] (Linnaeus, 1761)
      = *bioculatus* auctorum
    6. *scambus* Eaton, 1870
    7. *vernus*[2] Curtis, 1834 = *tenax* Eaton, 1870 **syn. nov.**
    8. *buceratus* Eaton, 1870
    9. *rhodani* (Pictet, 1844)
    10. *atrebatinus* Eaton, 1870
    11. *muticus*[3] (Linnaeus, 1758) = *pumilus* (Burmeister, 1839)
    12. *niger* (Linnaeus, 1761)
    13. *digitatus*[4] Bengtsson, 1912
  CENTROPTILUM Eaton, 1869
    14. *luteolum* (Müller, 1776)
    15. *pennulatum* Eaton, 1870
  CLOEON Leach, 1815
    16. *dipterum* (Linnaeus, 1758)
    17. *simile* Eaton, 1870
  PROCLOEON Bengtsson, 1915
    18. *pseudorufulum* Kimmins, 1957 = *rufulum* Eaton, nec Müller

Family HEPTAGENIIDAE = ECDYONURIDAE[5]
  RHITHROGENA Eaton, 1881
    19. *semicolorata* (Curtis, 1834)
    20. *haarupi* Esben-Petersen, 1909
  HEPTAGENIA Walsh, 1863
    21. *sulphurea* (Müller, 1776)
    22. *longicauda* (Stephens, 1836) = *flavipennis* (Dufour)
    23. *fuscogrisea* (Retzius, 1783) = *volitans* Eaton
    24. *lateralis* (Curtis, 1834)
  ARTHROPLEA Bengtsson, 1908 = HAPLOGENIA Blair
    25. *congener* Bengtsson, 1909 = *southi* Blair
  ECDYONURUS Eaton, 1868 = ECDYURUS Eaton
    26. *venosus* (Fabricius, 1775) = *forcipula*[6] (Pictet); Hincks & Dibb, 1935
    27. *torrentis* Kimmins, 1942
    28. *dispar* (Curtis, 1834) = *longicauda* Eaton
    29. *insignis* (Eaton, 1870)

Fam. LEPTOPHLEBIIDAE
LEPTOPHLEBIA Westwood, 1839
30. *marginata* (Linnaeus, 1758)
31. *vespertina* (Linnaeus, 1758)
PARALEPTOPHLEBIA Lestage, 1916
32. *submarginata* (Stephens, 1836)
33. *cincta* (Retzius, 1783)
34. *tumida* Bengtsson, 1930

HABROPHLEBIA Eaton, 1881
35. *fusca* (Curtis, 1834)

Family EPHEMERELLIDAE
EPHEMERELLA Walsh, 1862
36. *ignita* (Poda, 1761)
37. *notata* Eaton, 1887

Family POTAMANTHIDAE
POTAMANTHUS Pictet, 1844
38. *luteus* (Linnaeus, 1758)

Family EPHEMERIDAE
EPHEMERA Linnaeus, 1758
39. *vulgata* Linnaeus, 1758
40. *danica* Müller, 1764
41. *lineata* Eaton, 1870

Family CAENIDAE
BRACHYCERCUS Curtis, 1834
42. *harrisella* Curtis, 1834
CAENIS Stephens, 1836
43. *macrura* Stephens, 1836
    = *halterata* (Eaton nec Fabricius)
44. *moesta* Bengtsson, 1917
45. *robusta* Eaton, 1884
46. *horaria* (Linnaeus, 1758)
47. *rivulorum* Eaton, 1884

---

[1] The species identified by Eaton and others as *B. bioculatus* (L.) has been found to be *B. fuscatus* (L.), the true *bioculatus* (L.) belonging to another genus.

[2] I have long been doubtful whether *B. tenax* was really distinct from *B. vernus*. The characters given in previous editions of these keys were by no means satisfactory. There seemed to be some ecological differences, *vernus* occurring in slow lowland rivers and possibly preferring more alkaline waters, while *tenax* occurred in mountain streams at high altitudes. Frau Müller-Liebenau has been unable to separate the two species and Dr Macan is similarly unable to distinguish the nymphs. I am therefore placing *tenax* as a synonym of *vernus*.

[3] The species formerly known as *B. pumilus* (Burm.) has been re-named *B. muticus* (L.) in Müller-Liebenau's revision.

[4] Mr T. Gledhill has found in the River Frome in Dorset what may be an addition to the British list, *B. digitatus*. It is close to *B. niger*, but, whereas the last gill of *B. niger* is more or less symmetrical about the long axis, the inner margin of the last gill of B. *digitatus* runs straight for most of its length, then turns round to form a concavity before it reaches the tip. The differences between the adults are slight. Müller-Liebenau is doubtful whether these are two good species but decides to retain *B. digitatus* provisionally until more material is available.

[5] The family name Heptageniidae has priority over the name Ecdyonuridae and it has been changed accordingly.

[6] This species was provisionally placed on the British list by Hincks and Dibb (1935), on the evidence of nymphs resembling Schoenemund's figure of *E. forcipula* nymphs. Dr Rawlinson (1939) has shown that the pattern which Schoenemund claims to be typical of *E. forcipula* falls within the range of variation of *E. venosus* in this country. On the evidence available, the status of *E. forcipula* (Pictet) as a British species is very doubtful, and I have therefore dropped it from the British list.

COLLECTING AND PRESERVING

An ordinary insect net is required to collect Ephemeroptera; the mesh of the cloth need not be finer than the leno material usually supplied. It is an advantage to be able to attach to the net a moderately long handle for dealing with high-flying swarms. Most of the imagines taken in swarms will be males; the females and also the subimagines will more often be found during the day resting amongst foliage. Subimagines may be bred out to the imaginal stage if placed in a box with rough sides to which they can cling, and the moult is more likely to be successful if moistened blotting paper is placed in the box to maintain the humidity of the air. Subimagines for breeding out should be handled carefully to avoid damage to the wings. It is desirable to keep some subimagines for the collection, provided that they can be accurately associated with the imagines, as the pattern of the wings is in many species distinctive. These resting forms are most easily obtained by beating the branches of trees and shrubs near the water over the mouth of a net. Certain species, such as *Caenis* and *Potamanthus* are largely nocturnal and may come to light, and the former may also be found on the wing in the early morning soon after sunrise. Imagines of *Caenis* may also often be found trapped in spiders' webs near the water. A good hand lens or a low-power microscope, or preferably both, will be needed for identifying Ephemeroptera.

Ephemeroptera are very fragile creatures and are difficult to preserve in good condition. They may be pinned and preserved dry, either set or unset, or they may be kept in fluid. Dried specimens, particularly of the smaller species, suffer somewhat from shrivelling, but if necessary for study, the male genitalia may be examined by cutting off the apex of the abdomen and boiling it in dilute caustic potash solution (about 10%); after passing it through glacial acetic acid, it can be cleared in clove oil and then mounted in canada balsam on a cover glass and attached to the pin of the specimen. If the specimens are to be set, the wings should be spread out on a setting board as one would treat Lepidoptera, and in addition the fore legs and tails should be kept in place under the strips of setting paper. Strips of cellophane, as recommended by Mosely for

Trichoptera, are excellent for this purpose. When setting subimagines, it is necessary to cover the entire wings and also the tails with the setting-strips to prevent crumpling and shrinking in drying. Well-set Ephemeroptera make attractive cabinet specimens, but they unfortunately take up a great deal of space, and it is convenient to keep some specimens pinned through the side of the thorax, with the wings folded over the back. Another method is to attach the specimen with wings outspread to a strip of celluloid by means of gum tragacanth. Specimens so mounted are less fragile but not quite so easy to examine. Creosote should never be placed as a preservative in store boxes containing celluloid, as it will rapidly discolour the strips to a rich red-brown.

Ephemeroptera for pinning may be killed in a cyanide bottle, by exposure to the fumes of chopped and crushed laurel leaves, or to the fumes of strong ammonia. Many specimens should not be placed in the bottle at one time, and as far as possible they should be protected from damage due to shaking about by placing loosely-crumpled tissue paper in the bottle. Specimens should not be allowed to remain in the killing bottle for more than a few hours, particularly when cyanide is used, or they may acquire a pinkish tinge which spoils them for the collection. Specimens for the fluid collection may be placed directly into alcohol of about 70% strength and after an hour or so transferred to a dilute solution of formaldehyde, one part of 40% formaldehyde in 19 parts of water. The preliminary bath of alcohol is necessary to wet the specimens thoroughly, as owing to the density of the formaldehyde solution, the specimens will not sink if placed into it direct, nor will the solution penetrate the tissues and preserve them from decomposition. Specimens intended for mounting as miscroscope preparations should be retained in 70% alcohol, as less boiling in potash solution will then be required than in the case of formalin material. Nymphs may be placed directly into the formaldehyde solution. Examples in fluid must also be protected against undue shaking, or it will be found that the legs and gills become detached. Tubes containing specimens for transit should always be *completely* filled with liquid. An air-bubble running back and forth in a tube can do almost as much damage as a small pebble.

Whatever method of preservation is adopted, it is essential if the collection is to have any scientific value that specimens should have adequate locality labels. The county should always be quoted, as many place names are duplicated in other parts of the country. Any special ecological data, such as type of habitat, time of swarming, etc., should preferably be placed on a separate label. Above all, the label should be legibly written, in waterproof ink for fluid specimens, and abbreviations, which will probably be unintelligible to other workers, should be avoided. A collection which was presented to the British Museum some years ago had its value much reduced by its locality labels, which were often abbreviated to initial letters, as WX, LM, G. Oak, etc. The practice of numbering specimens and keeping all relevant data in a note book or register is equally undesirable, inasmuch as time is wasted in consulting the register, and should the register be lost, the collection is largely useless.

FISHERMEN'S NAMES

Nymphs and adults of the commoner species of Ephemeroptera are an important source of food for fishes. Hatching nymphs, subimagines and imagines on the surface of the water are frequently taken by trout, and in consequence have been selected as patterns for artificial flies and given popular names by anglers. These popular names may apply to a particular species or to a group of species either closely allied or closely resembling each other. In general the subimagines are known as *duns* and the imagines as *spinners*. To avoid confusion it is suggested that the term "mayfly" be restricted to the genus *Ephemera*, and that as a comprehensive vernacular name for the order the term "day-flies" be employed, thus falling into line with the German entomologists, who use the name "Eintagsfliegen." The following names are in more or less general use (see also Harris 1952):

Angler's Curse	*Caenis* spp.
August Dun	*Ecdyonurus dispar* subimago
Autumn Dun	*Ecdyonurus dispar* subimago
Black Drake	*Ephemera danica* male imago
Blue Dun (Ronalds)	*Baetis vernus* subimago
Blue Winged Olive Dun (B.W.O.)	*Ephemerella ignita* subimago
Blue Winged Pale Watery Dun	*Centroptilum pennulatum* subimago
Brown May Dun	*Heptagenia fuscogrisea* subimago
Claret Dun	*Leptophlebia vespertina* subimago
Dark Dun	*Heptagenia lateralis* subimago
Dark Mackerel (Ronalds)	*Ephemera vulgata* imago
Dark Olive Dun	*Baetis atrebatinus* subimago
Dun Drake (Ronalds)	*Ecdyonurus venosus* subimago
False March Brown	*Ecdyonurus venosus* subimago
Great Red Spinner	*Rhithrogena haarupi, Ecdyonurus venosus, E. torrentis, E. dispar* imagines

Green Drake	*Ephemera danica* subimago
Grey Drake	*Ephemera danica* female imago
Iron Blue Dun	*Baetis muticus, B. niger* subimagines
Jenny Spinner	*Baetis muticus, B. niger* male imagines
July Dun (Ronalds)	Probably *Ephemerella ignita* subimago
July Dun	*Baetis scambus* subimago
Lake Olive Dun	*Cloeon simile* subimago
Large Amber Spinner	*Centroptilum pennulatum* female imago
Large Claret Spinner	*Leptophlebia vespertina* female imago
Large Dark Olive Dun	*Baetis rhodani* subimago
Large Dark Olive Spinner	*Baetis rhodani* imago
Large Green Dun	*Ecdyonurus insignis* subimago
Large Green Spinner	*Ecdyonurus insignis* imago
Large Summer Spur-wing	*Centroptilum pennulatum* imago
Late March Brown	*Ecdyonurus venosus* subimago
Little Amber Spinner	*Centroptilum luteolum* female imago
Little Claret Spinner	*Baetis muticus, B. niger* female imagines
Little Pale Blue Dun (Ronalds)	*Procloeon pseudorufulum* subimago
Little Sky-blue Dun	*Centroptilum luteolum* subimago
Little Yellow May Dun	*Heptagenia sulphurea* subimago
March Brown	*Rhithrogena haarupi* subimago
Mayfly	*Ephemera* spp., more particulary *E. danica*
Medium Olive Dun	*Baetis vernus* subimago
Olive Duns	*Baetis rhodani, B. vernus, B. atrebatinus* and *B. scambus* subimagines
Olive Upright	*Rhithrogena semicolorata* subimago
Pale Evening Dun	*Procloeon pseudorufulum* subimago
Pale Watery Duns	*Baetis fuscatus, Centroptilum luteolum* and *Procloeon pseudorufulum* subimagines
Pond Olive Dun	*Cloeon dipterum* subimago
Purple Dun	*Paraleptophlebia cincta* subimago
Purple Spinner	*Paraleptophlebia cincta* imago
Red Spinner (Ronalds)	*Ecdyonurus dispar* imago
Red Spinner	*Baetis rhodani* female imago
Sherry Spinner	*Ephemerella ignita* imago
Small Dark Olive Dun	*Baetis scambus* subimago

Spent Gnat	*Ephemera danica* imago floating on the water with wings outspread after oviposition
Summer Mayfly	*Siphlonurus* spp.
Turkey Brown	*Paraleptophlebia submarginata* subimago
Whirling Blue Dun (Ronalds)	*Ecdyonurus dispar* subimago
White Midge	*Caenis* spp.
Yellow Dun (Ronalds)	*Ecdyonurus dispar* subimago
Yellow Evening Dun	*Ephemerella notata* subimago
Yellow Evening Spinner	*Ephemerella notata* imago
Yellow Hawk	*Heptagenia sulphurea* subimago
Yellow May Dun	*Heptagenia sulphurea* subimago
Yellow Upright	*Rhithrogena semicolorata* imago

KEY TO FAMILIES

This key applies to both imagines and subimagines. Except in the Caenidae the wings of imagines are not fringed with hairs. In subimagines the wings are fringed (fig. 2), the colours are generally duller, and the male genitalia are not fully developed.

1 M and Cu_1 in fore wing strongly divergent at base (figs. 1, 5). Large (fore wing 12–24 mm long). Three long filaments at hind end—   **2**

— M and Cu_1 in fore wing subparallel at base (fig. 6). Small to large. Two or three terminal filaments—   **3**

2 Wings more or less marked with brown. Abdomen light with brown markings. A_1 in fore wing simple (fig. 1)—
    EPHEMERIDAE, p. 64

— Wings unspotted, yellow. Abdomen yellowish with darker dorsal stripe. A_1 in fore wing forked (fig. 26)—
    POTAMANTHIDAE, p. 63

3 Hind tarsus with five free segments (fig. 25). Medium-sized or large. With three long filaments at hind end—
    HEPTAGENIIDAE, p. 42

— Hind tarsus with four or three free segments, the basal segment(s) being more or less fused to the tibia (fig. 5)—   **4**

4 Hind wings absent, fore wings milky, margins in imago fringed, outer fork of Rs very deep (fig. 29). Small, forewing 3–7 mm long. With three long filaments at hind end—
    CAENIDAE, p. 69

— Hind wings present or absent, wings hyaline, in imago not fringed, outer fork of Rs normal or detached basally—   **5**

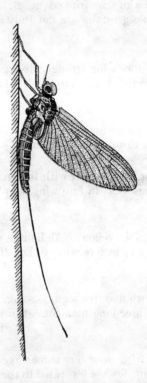

Fig. 2. *Baetis rhodani*. Female subimago in resting position × 3·5.

5 R_5 of fore wing detached basally from R_4 (fig. 6). Hind tarsi apparently of three segments. Small or medium-sized. Two long filaments at hind end— BAETIDAE, p. 29

— R_4 and R_5 in fore wing forked normally. Tarsi apparently of 4 segments— 6

6 R_4 and R_5 in hind wing separate at wing margin (fig. 3). Two long filaments at hind end. Medium-sized or large— SIPHLONURIDAE, p. 25

— R_4 and R_5 fused in hind wing. Three long filaments at hind end. Medium-sized— 7

7 In fore wing Cu_2 either nearer to A_1 at base or at most midway between Cu_1 and A_1 (figs. 21, 23)— LEPTOPHLEBIIDAE, p. 55

— In fore wing Cu_2 nearer to Cu_1 than A_1 at base (fig. 25)— EPHEMERELLIDAE, p. 61

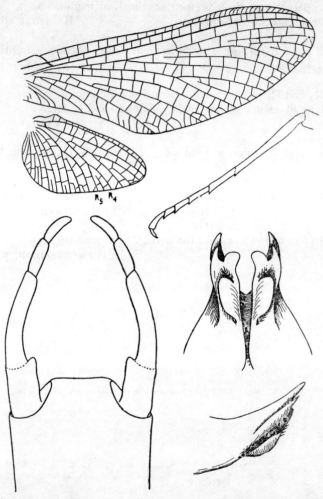

Fig. 3. *Siphlonurus lacustris*, male. Wings (× 6·5), hind tibia and tarsus, claspers (ventral) and penis-lobes (ventral and lateral).

KEYS TO GENERA AND SPECIES

Family SIPHLONURIDAE

1 Hind tarsus about one-and-a-half times as long as tibia, claws similar (fig. 3)— SIPHLONURUS Eaton.

— Hind tarsus slightly shorter than tibia, claws dissimilar (fig. 5)— AMELETUS Eaton.

Genus SIPHLONURUS Eaton

IMAGINES

1 Femora with a dark reddish brown transverse band on outer surface before apex (fig. 4L)— **Siphlonurus linnaeanus** (Etn.).

Limestone lakes and slow, deep river pools in Ireland, River Tummel and River Cree in Scotland. May to August. Not recorded from the Lake District.

— Femora unbanded— 2

2 Hind angles of tergite IX only slightly produced, VIII scarcely widened behind; lobes of penis with acute apices (fig. 3)— **Siphlonurus lacustris** Etn.

Not uncommon in lakes, slow streams and high tarns, May to September.
   Lake District, lakes, slow streams and high tarns up to an altitude of 2,500 ft.

— Hind angles of tergite IX strongly produced behind to form pointed spines; VIII tergite behind, and IX tergite noticeably widened; lobes of penis rounded at apices (fig. 4A)—
Siphlonurus armatus Etn.

Few records; lakes, ponds and slow streams, May, June and August.
Not recorded from the Lake District, but occurring in the River Winster, just outside the boundaries.

SUBIMAGINES

1 Wings greyish, hind wings with the apices of the veins and the posterior margin whitish, giving the appearance of a pale border— **Siphlonurus linnaeanus** (Etn.)

— Hind wings without a pale border— 2

2 Wings brownish grey, with cross-veins clearly visible; hind angle of IX tergite produced as in imago—
Siphlonurus armatus Etn.

— Wings uniform greenish grey, main veins darker, cross-veins not very definite— **Siphlonurus lacustris** Etn.

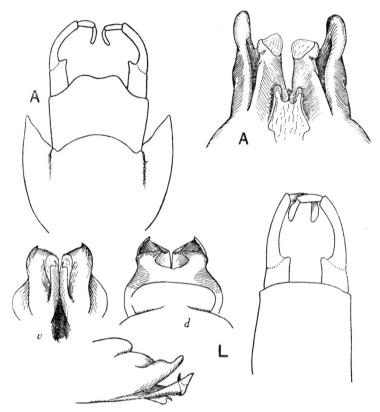

Fig. 4. *Siphlonurus*, male. A, *S. armatus*, claspers (ventral) and penis-lobes (dorsal); L, *S. linnaeanus*, claspers (ventral) and penis-lobes (dorsal, ventral and lateral).

Genus AMELETUS Eaton

Only one species is recorded from the British Isles, **Ameletus inopinatus** Etn.

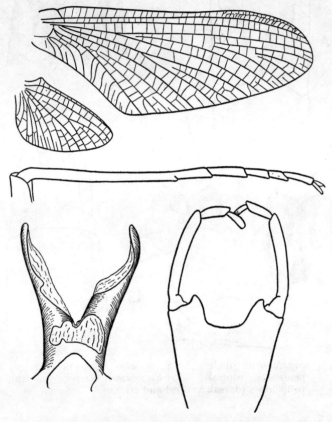

Fig. 5. *Ameletus inopinatus*, male. Wings (× 7·5), hind tibia and tarsus, penis-lobes (dorsal), and claspers (ventral).

Dark brown in colour, wings hyaline, cross-veins rather weak; lobes of penis slender and divergent, with a deep excision between them (fig. 5).

Wings of subimago reddish or yellowish brown.

Mountain and hill streams, June, early July.
Lake District, occurs in high streams such as Rydal Head, 1,800 ft., Stock Ghyll Head, 1,700 ft., Whelpside Ghyll, about 2,000 ft.

Family BAETIDAE

1. Hind wing present, but sometimes small and narrow (figs. 6, 11)—   2

— Hind wing absent—   3

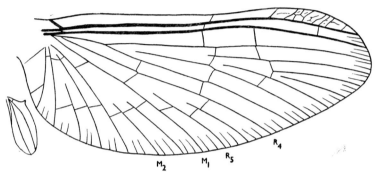

Fig. 6. *Baetis scambus.* Male wings × 13.

2. Marginal intercalaries in fore wing paired (fig. 6)—   BAETIS Leach.

— Marginal intercalaries in fore wing single (fig. 11)—   CENTROPTILUM Eaton.

3. First cross-vein between R_1 and R_2 in line with the one below it; hind tarsus with the first segment three times as long as the second (fig. 13)—   PROCLOEON Bengtsson.

— First cross-vein between R_1 and R_2 not in line with the one below it; hind tarsus with the first segment twice as long as second (fig. 12D)—   CLOEON Leach.

Genus BAETIS Leach

MALE IMAGINES

The present key to species is not entirely satisfactory in some respects, but the identification of the various species of *Baetis* presents considerable difficulties. Even in the males, for which this key is designed, the only characters available are often comparative ones rather than alternatives. The male forceps do not offer very tangible characters and the venation of the hind wing is somewhat unstable, and in some cases differs on opposite sides of the same individual.

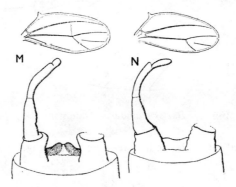

Fig. 7. *Baetis*. Male hind wings and genitalia (ventral). M, *B. muticus*; N, *B. niger*.

1 Second longitudinal vein in hind wing forked (fig. 7); forceps from side not arched— 2

— Second longitudinal vein in hind wing (figs. 8, 9) not forked (rarely forked in *buceratus*, the forceps of which are arched from the side)— 4

2 Hind wing with three veins, membrane somewhat milky hyaline; male abdomen with tergites II-VII white, VIII-X pitch-brown; terminal segment of forceps very short (fig. 7M)—
Baetis muticus (L.)

Common in rivers and small streams. April to September.
Lake District, common in rivers and small streams up to at least 1,500 ft.

— Hind wing with two longitudinal veins, membrane vitreous; abdomen as in *muticus*, terminal segment of forceps long (fig. 7N)— 3

3 Forceps: basal segment cylindrical or tapering; second and third segments with scarcely any constriction at the point where they meet; fourth segment about half as long as the third (fig. 7N)— **Baetis niger** (L.)

Rivers and streams with vegetation. May to September.
Lake District, Hog House Beck, Cunsey Beck.

— Forceps: basal segment tapering; shaft constricted in the region where second and third segments meet; fourth segment about two thirds as long as third— **Baetis digitatus** Bengt.

Rivers and streams with vegetation. May to September.
Not recorded from the Lake District.

4 (1) Hind wing with a costal process near the base— 5

— Hind wing without a costal process; colouring much as in *B. rhodani*; an acute chitinised process between the bases of the forceps (fig. 8A)— **Baetis atrebatinus** Etn.

Rather scarce or local, in alkaline rivers and streams. May to October.
Not recorded from the Lake District.

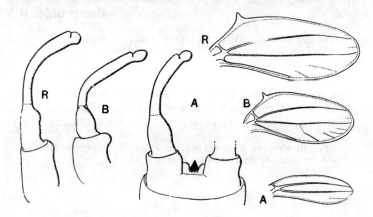

Fig. 8. *Baetis*. Male hind wings and genitalia (ventral). R, *B. rhodani*; B, *B. buceratus*; A, *B. atrebatinus*.

5 Forceps base with a short cylindrical truncate process near the inner distal corner— 6

— No such process on forceps base— 8

6 A small point on inner margin near apex of second segment of forceps (fig. 9V). Turbinate eyes, large oval, more convex towards the outer than the inner margin—
Baetis vernus Curt.
(**B. tenax** Etn.)

Moderately common in sluggish lowland rivers, May to September, and common in small streams above about 1,500 feet (*B. tenax*), April to September.
Lake District, occurs in becks above 1,500 feet.

— Second segment of forceps unarmed— 7

7 Turbinate eyes lemon-yellow; forceps fig. 9B—
Baetis fuscatus (L.)

Common in rivers on sand and gravel and in vegetation, possibly with a preference for alkaline waters. May to September. Not recorded from the Lake District.

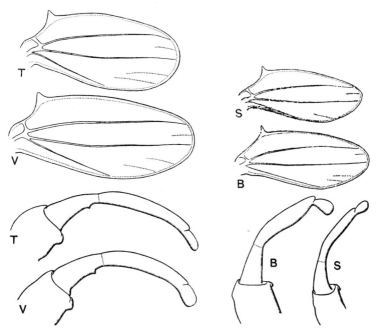

Fig. 9. *Baetis*. Male hind wings and genitalia (ventral). T, *B. tenax*; V, *B. vernus*; B, *B. fuscatus*; S, *B. scambus*.

— Turbinate eyes sepia-brown; forceps fig. 9S—
Baetis scambus Etn.

As preceding, possibly more often in soft water. February to November.

Lake District, rivers such as the Leven, Rothay and Troutbeck.

8 (5) A swelling near apex of inner margin of forceps base; first segment generally almost parallel-sided, sometimes the inner margin concave. Faceted surface of turbinate eyes liver brown with a lighter ring round the margin, shaft also lighter and encircled by dark rings. Inner margin of second segment of forceps somewhat convex— **Baetis rhodani** (Pict.)

Common in the more rapid parts of rivers and in small streams. The imago is commonest in spring and late autumn, but has been taken in most months of the year.

Lake District, common in rivers and small, stony streams up to at least 1,000 ft.

— A large swelling near apex of inner margin of forceps base. First segment of forceps pointing upwards, second bent so that distal portion points downwards (fig. 8B)—
Baetis buceratus Etn.

Not common, in rivers, June and September.
Not recorded from the Lake District.

SUBIMAGINES

The subimagines of *Baetis* are at least as difficult to identify as the adults, and I have not attempted to make a key. *B. muticus* and *B. niger* can be distinguished by their colour and the venation of the hind wing. *B. rhodani* has a distinctive pattern on the metatergum; this pattern pertains to the subimaginal skin and can often be seen showing through the skin of the mature nymph.

B. muticus. Wings charcoal-black.
B. niger and **B. digitatus.** Wings blackish grey.
B. atrebatinus. Wings much as in *rhodani*, but without costal process to hind wing.
B. buceratus. Fore wing smoky, hind wing paler, setae piceous.

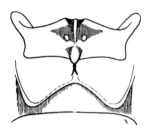

Fig. 10. *Baetis rhodani* (Pict.). Pattern of metatergum of subimago.

B. rhodani. Wings ash-grey or dark grey-brown with greenish grey veins. A distinctive pattern to the metatergum (fig. 10).
B. vernus. Wings and setae smoky grey.
B. fuscatus. Fore wing grey-brown or smoky grey, hind wing paler, setae pale grey.
B. scambus. Wings and setae cinereous.

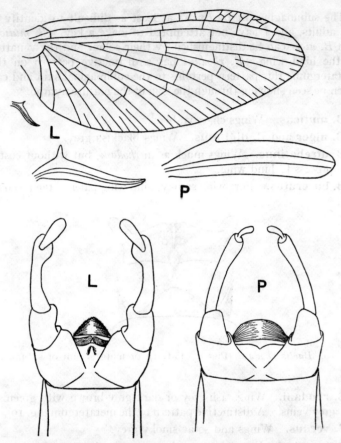

Fig. 11. *Centroptilum*. Male wings (× 10), enlarged hind wings, and genitalia (ventral) of L, *C. luteolum*; P, *C. pennulatum*.

Genus CENTROPTILUM Eaton

IMAGINES

1 Hind wing acute at apex; male abdominal segments II-VII translucent whitish; terminal segments of forceps large (fig. 11L)— **Centroptilum luteolum** (Müll.)

Common in rivers and streams, with moderate to slow flow and on the stony shores of lakes. April to November.
Lake District, common in rivers, streams and along the shores of lakes such as Windermere.

— Hind wing rounded at apex; male abdominal segments II-VI translucent whitish, apical margins narrowly reddish orange; terminal segment of forceps small (fig. 11P)— **Centroptilum pennulatum** Etn.

In the same sort of places as *C. luteolum* but less widespread. May to October.
Lake District, Hog House Beck, near Wray Castle, a few. (T. T. Macan).

SUBIMAGINES

1 Wings ashy grey, small species (fore wing 6-7 mm)— **Centroptilum luteolum** (Müll.)

— Wings blue-grey, large species (fore wing 9 mm)— **Centroptilum pennulatum** Etn.

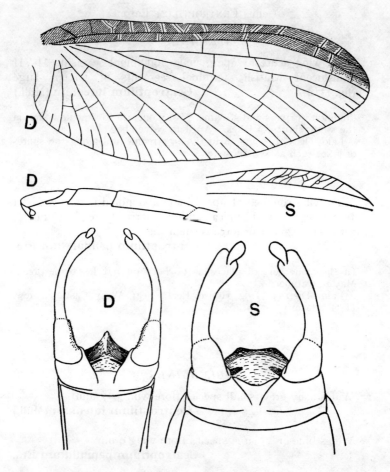

Fig. 12. *Cloeon*. D, *C. dipterum*, female wing (× 9), male hind tarsus, and male genitalia (ventral); S, *C. simile*, pterostigmatic region and male genitalia (ventral).

Genus CLOEON Leach

IMAGINES

1 Pterostigma with three to five cross-veins; costal and sub-costal areas in female yellowish brown, with pale areas bordering the cross-veins. Forceps slender, penis in form of a trangular plate (fig. 12D)— **Cloeon dipterum** (L.)

Common in productive ponds; occurred abundantly in static water tanks in London during the 1939-45 War. May to October.
Lake District, common in small ponds.

— Pterostigma with nine to eleven cross-veins; costal and sub-costal areas in female uncoloured. Forceps stouter, penis trapezoidal (fig. 12S)— **Cloeon simile** Etn.

Common in ponds, lakes and slow rivers, often in vegetation in deep water. March to November.
Lake District, occurring in tarns, lakes, slow streams and rivers.

SUBIMAGINES

Wings light blackish grey; three to five pterostigmatic cross-veins— **Cloeon dipterum** (L).

— Wings mouse-grey, tinged with yellowish at base and along costa; nine to eleven pterostigmatic cross-veins— **Cloeon simile** Etn.

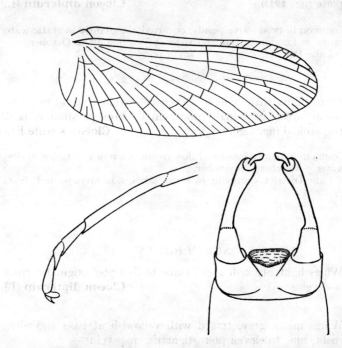

Fig. 13. *Procloeon pseudorufulum*. Wing (× 10), hind tarsus and male genitalia (ventral).

Genus PROCLOEON Bengtsson

One species, **Procloeon pseudorufulum** Kimmins, 1957. The name was proposed to replace *P. rufulum* (Eaton) nec Müller.

The venational character given in the generic key is not entirely stable, but the relative proportions of the hind tarsal segments are constant. The imago certainly differs in general appearance from *C. dipterum* and *C. simile*, having more the colouring of *Baetis fuscatus* and *Centroptilum luteolum*.

The wings of the subimago are light greyish or greyish white, often tinged with greenish along the main veins; six to eight pterostigmatic cross-veins.

Common in Ireland, less common in England, Scotland and Wales. Occurs in slow flowing rivers. Adult crepuscular in habits. May to October.

Lake District, R. Brathay, fair numbers July 1950 and 1951 (T. T. Macan).

Family HEPTAGENIIDAE

1 First four segments of male fore tarsus subequal, about twice as long as fifth; R_{4+5} in hind wing unforked (fig. 17)—
ARTHROPLEA Bengtsson

— First segment of male fore tarsus shorter than one or other of the succeeding segments, R_{4+5} in hind wing forked (fig. 15)— 2

2 Penis-lobes outspread, boot-shaped (fig. 18)—
ECDYONURUS Eaton

— Penis-lobes not outspread and boot-shaped— 3

3 Penis-lobes contiguous, slightly dilated or egg-shaped (figs. 15, 16)— HEPTAGENIA Walsh

— Penis-lobes separated by a wide, deep U-shaped excision, cylindrical (fig. 14)— RHITHROGENA Eaton

Genus RHITHROGENA Eaton

IMAGINES

1 Small species, expanse 18–24 mm. Wings in basal half more or less tinted with pale golden brown; outer apical angles of penis-lobes from beneath acute (fig. 14S)—
Rhithrogena semicolorata (Curt.)

Common in fast, stony rivers and streams. April to September. Lake District, common in rivers and becks.

— Robust species, expanse 29–38 mm. Wings very indistinctly shaded with brownish at base; outer apical angle of penis-lobes from beneath rounded (fig. 14H)—
Rhithrogena haarupi Esb.-Pet.

Common but local on large, rapid rivers. Late March to early May. Not recorded from the Lake District.

SUBIMAGINES

1 Wings pale mouse-grey, hind wings paler, cross-veins not bordered— **Rhithrogena semicolorata** (Curt.)

— Wings pale yellowish grey, cross-veins bordered with blackish—
Rhithrogena haarupi Esb.-Pet.

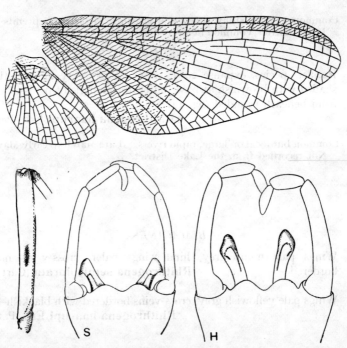

Fig. 14. *Rhithrogena*. S, *R. semicolorata*, wings (× 7·5), femoral markings, and male genitalia (ventral). H, *R. haarupi*.

Genus HEPTAGENIA Walsh

IMAGINES

1 Hind tarsus with basal segment shorter than second, forceps-base simple (fig. 15)—   2

— Hind tarsus with basal segment longer than second, forceps-base toothed (fig. 16)—   3

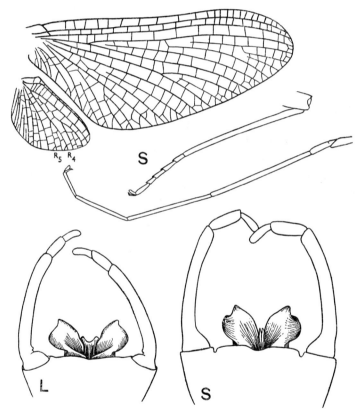

Fig. 15. *Heptagenia*, male. Wings (× 8), front tarsus, hind tibia and tarsus, and genitalia (ventral). S, *H. sulphurea*; L, *H. longicauda*.

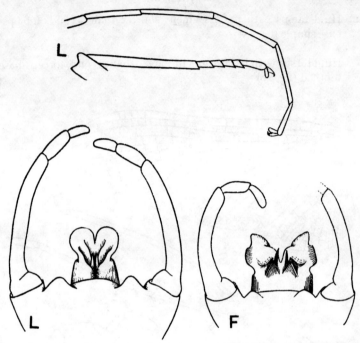

Fig. 16. *Heptagenia*, male. Fore tarsus, hind tibia and tarsus, and genitalia (ventral). L, *H. lateralis*; F, *H. fuscogrisea*.

2 Femora, at least the anterior, with two flesh-coloured rings, one midway, one at apex; a small black spot on the side of the metathorax, above the hind coxa. General colour yellowish brown, abdomen in male with segments II-VIII translucent; wings faintly suffused with yellowish, stronger in costal and subcostal areas (fig. 15L)—
(*H. flavipennis* Duf.) **Heptagenia longicauda** (Steph.)

Very few British records. Old records in rivers such as the Thames and the upper part of the Wey, late May and early June. The duns emerge after sunset.
Not recorded from the Lake District.

— Femora not so marked, at most only apex darker; a short line just behind the fore coxa and one to three dots above the mid coxa black; general colour light yellowish brown in male, yellow in female; subcostal and costal areas often yellowish (fig. 15S)— **Heptagenia sulphurea** (Müll.)

Common in the lower course of rivers and in rocky, limestone lakes. May to August.
Lake District, common in such rivers as the Brathay, Rothay and Leven.

3 (1) Penis-lobes egg-shaped; general colour dark brown, with a bright yellow streak on thorax directed forwards from the fore wing-base (fig. 16L); no sherry-coloured bands on femora—
Heptagenia lateralis (Curt.)

Common in rapid streams and along the stony shores of large lakes. May to September.
Lake District, very common in becks and along the stony shores of lakes such as Windermere, Ullswater, etc.

— Penis-lobes with an excision on outer margin near apex; general colour dark brown, no bright yellow thoracic streak (fig. 16F); two sherry-coloured bands on each femur—
(*H. volitans* Etn.) **Heptagenia fuscogrisea** (Retz.)

Common in rocky, limestone lakes and rivers in Ireland, local and uncommon in England. May and early June.
Not recorded from the Lake District.

SUBIMAGINES

1 Wings yellow or greenish yellow—   2

— Wings grey or greyish yellow—   3

2 Femora marked with flesh-coloured bands and side of thorax with a black dot as in imago—
Heptagenia longicauda (Steph.)

— Femora unmarked, sides of thorax marked with black as in imago—   **Heptagenia sulphurea** (Müll.)

3 (1) Wings grey, sometimes faintly marked with transverse bands; no sherry-coloured bands on femora—
Heptagenia lateralis (Curt.)

— Wings greyish yellow, veins and cross-veins darker; two sherry-coloured bands on each femur—
Heptagenia fuscogrisea (Retz.).

Genus ARTHROPLEA Bengtsson (=*Haplogenia* Blair)

Only one species is recorded from the British Isles, **Arthroplea congener** Bengtsson (=*Haplogenia southi* Blair). Dark brown in colour, wings hyaline, very slightly infuscate in basal half. Penis-lobes adjacent, evenly expanded towards apex, provided with a long spine on each side at base (fig. 17).

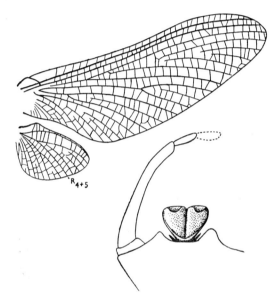

Fig. 17. *Arthroplea congener*, male. Wings ($\times$ 6) and genitalia (ventral). (After Blair.)

The subimago is unknown to me.

The only British specimen known to me was taken at Stanmore, Middlesex, in June 1920.

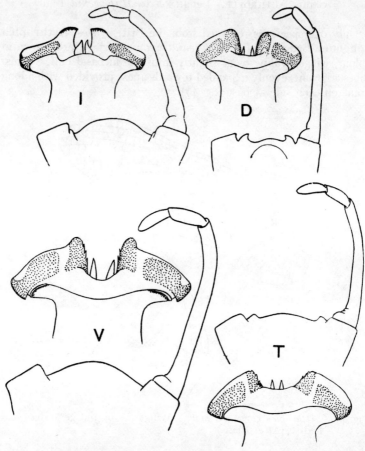

Fig. 18. *Ecdyonurus*, male. Forceps-base, right clasper and penis-lobes (dorsal, more enlarged). I, *E. insignis*; D, *E. dispar*; V, *E. venosus*; T, *E. torrentis*.

Genus ECDYONURUS Eaton

MALE IMAGINES

1 Sternites II-VIII with three black lines and two black spots on a yellow ground (fig. 19); forceps-base simple, inner angles of penis-lobes incurved (fig. 18I)— **Ecdyonurus insignis** (Etn.)

Rather more local than the other species of *Ecdyonurus*. It occurs in rather large, fast streams and rivers, possibly with a preference for alkaline waters. May to September.
Not recorded for the Lake District, but occurs in the River Eden.

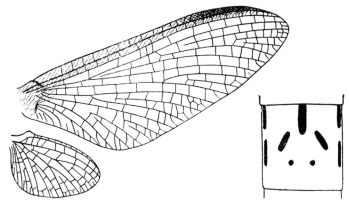

Fig. 19. *Ecdyonurus*. *E. venosus*, wings (× 6); *E. insignis*, sternal pattern of abdomen.

— Sternites not so marked, ground colour reddish or brownish— 2

2 Forceps-base untoothed; proportions of male fore tibia to tarsus 1:1·5 (figs. 18V, 19)— **Ecdyonurus venosus** (Fabr.) (*E. forcipula* (Pictet))

Common in rapid, stony rivers. May to October.
Lake District, common in such rivers as the Rothay and Troutbeck.

— Forceps-base more or less toothed; proportions of male fore tibia to tarsus 1:1·85— 3

3 Forceps-base moderately toothed, teeth not incurved; penis-lobes boot-shaped (fig. 18T)— **Ecdyonurus torrentis** Kim.

As abundant as *venosus*, but with a preference for smaller streams. March to July.
Lake District, common in becks.

— Forceps-base strongly toothed, teeth incurved; penis-lobes subtriangular (fig. 18D)— **Ecdyonurus dispar** (Curt.) (*E. longicauda* Etn., nec Steph.)

Stony rivers, particularly in the autumn, also large lakes. June to October.
Lake District, common along the stony shores of Windermere, Ullswater, etc., and in rivers such as the Crake and Troutbeck.

SUBIMAGINES

1. Sternites marked with black as in imago (fig. 19)—
   Ecdyonurus insignis (Etn.)
— Sternites not so marked— 2

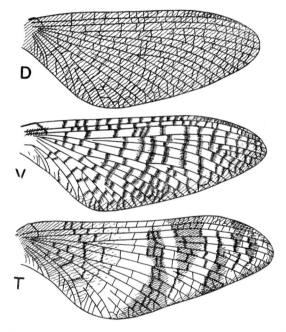

Fig. 20. *Ecdyonurus*. Fore wings of subimagines (× 5). D, *E. dispar*; V, *E. venosus*; T, *E. torrentis*.

2. Wing appearing uniformly greyish yellow, cross-veins only very finely bordered with blackish (fig. 20D)—
   Ecdyonurus dispar (Curt.)
— Cross-veins strongly bordered with blackish, giving a mottled or banded appearance— 3

3. Wing mottled with blackish (fig. 20V)—
   Ecdyonurus venosus (Fabr.)
— Wing with more or less definite transverse bands of blackish (fig. 20T)— **Ecdyonurus torrentis** Kim.

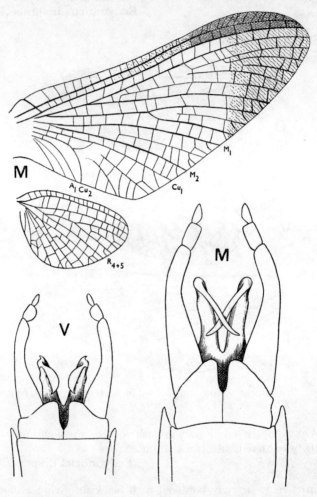

Fig. 21. *Leptophlebia*. Wings (× 6·5) and male genitalia (ventral). M, *L. marginata*; V, *L. vespertina*.

Fam. LEPTOPHLEBIIDAE

1 Costal margin of hind wing smoothly rounded; costal area long and narrow (figs. 21, 23)—   2

— Costal margin of hind wing with a strong projection about midway, costal area with its basal half very broad (fig. 24)—
HABROPHLEBIA Etn.

2 Cu_2 in fore wing at base lies mid-way between Cu_1 and A_1 (fig. 21)—   LEPTOPHLEBIA Westw.

— Cu_2 in fore wing at base lies nearer to A_1 than Cu_1 (fig. 23)—
PARALEPTOPHLEBIA Lest.

Genus LEPTOPHLEBIA Westwood

IMAGINES

1 Fore wing smoky brownish, particularly towards apex, at least the costal and subcostal areas around the pterostigma brown; veins brownish yellow or light piceous; penis-lobes at apices rounded, each with a recurved finger-like process ventrally (fig. 21M)— **Leptophlebia marginata** (L.)

Moderately abundant in lakes and ponds, slow streams and the slower parts of small, stony streams, sometimes at a considerable altitude (over 2,500 ft.). April to mid-June.
Lake District, common along the shores of lakes and in tarns, also occurring in small, stony streams.

— Fore wing entirely hyaline; venation colourless, except C, Sc and R_1, which are brownish; penis-lobes with a small hook apically, recurved ventral process blade-like (fig. 21V)—
Leptophlebia vespertina (L.)

Common in lakes, tarns and also occurs in small, stony streams, favouring an acid or peaty bottom. April to August.
Lake District, common in the shallower parts of lakes such as Windermere and Esthwaite Water, also in ponds.

Genus LEPTOPHLEBIA

SUBIMAGINES

1 Both wings brownish grey in life, hind wing scarcely paler than fore wing; cross-veins distinctly margined with brownish—
Leptophlebia marginata (L.)

— Fore wing in life mouse- or blue-grey, hind wing very pale almost buff: cross-veins not noticeably margined—
Leptophlebia vespertina (L.)

> Specimens of *L. vespertina* in collections tend to fade to a brownish grey, and in consequence the contrast between the fore and hind wings is considerably lessened.

Genus PARALEPTOPHLEBIA Lestage

IMAGINES

1 Setae white in male, yellowish in female; male abdomen usually with segments II-VII whitish, translucent, sometimes pale yellowish brown, VIII-X piceous. Penis-lobes with a broad short excision between the apices, which are rounded and carry a small blunt projection externally and a pair of recurved, pointed, blade-like processes ventrally, sometimes curving outward (fig. 22C)— **Paraleptophlebia cincta** (Brauer)

> Common in streams and small, rather fast rivers of an alkaline nature. May to August.
> Not recorded from the Lake District.

— Setae yellowish brown or brownish— 2

2 Male with segments III-VI yellowish brown, translucent, longitudinal veins yellow-brown; wings of female pale smoky brown. Penis-lobes with a deep excision between apices, which are outspread, and slender; about midway on lower surface of each lobe are two acute spines, the outer the longer (fig. 22T)— **Paraleptophlebia tumida** Bengtss.

> Recorded from two localities in Hampshire, where the nymphs occur in chalk-fed streams containing much aquatic vegetation, and from a small, weedy, sluggish stream, ceasing to flow in summer, in Cambridgeshire. May.

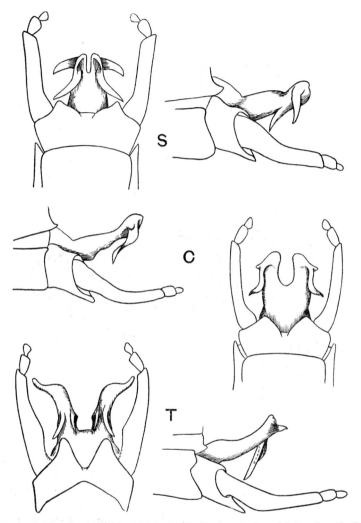

Fig. 22. *Paraleptophlebia*. Male genitalia (ventral and lateral). S, *P. submarginata*; C, *P. cincta*; T, *P. tumida*.

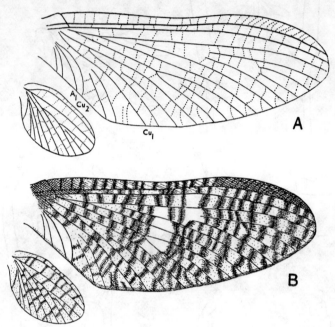

Fig. 23. *Paraleptophlebia submarginata*, male. Wings of A, imago, B, subimago × 7·5.

— Male with segments III-VI dark brown, paler at extreme base, longitudinal veins (except C, Sc, and R_1) usually white (fig. 23A); wings of female colourless. Penis-lobes with a short narrow excision apically, each apex armed with a pair of outwardly and downwardly directed spines (fig. 22S)—
Paraleptophlebia submarginata (Steph.)

Common in small, moderately fast streams. May to July.
Lake District, moderately common in small, stony streams.

Genus PARALEPTOPHLEBIA Lestage

SUBIMAGINES

1 Wings uniformly blackish grey—
   Paraleptophlebia cincta (Brauer)

— Wings uniformly mouse-grey, paler than in *cincta*—
   Paraleptophlebia tumida Bengtss.

— Wing membrane pale fawn, cross-veins heavily bordered with blackish, with a pale area in the centre of the fore-wing (fig. 23B)— **Paraleptophlebia submarginata** (Steph.)

Genus HABROPHLEBIA Eaton

Only one species, **Habrophlebia fusca** (Curt.), has been recorded from the British Isles.

Wings of the imago vitreous, longitudinal nervures and cross-veins of pterostigmatic region brownish. Abdomen brownish-piceous, segments II-VII translucent, setae light brownish grey. Penis-lobes slender, divergent, each with a slender basally-directed spine ventrally (fig. 24).

Wings of subimago blackish grey, with dark brown veins.

> Slow productive streams with plenty of aquatic vegetation, usually fairly abundant. May to September.
> Lake District, recorded from slow streams such as Hog House Beck and the upper reach of Nor Moss Beck, near Wray Castle.

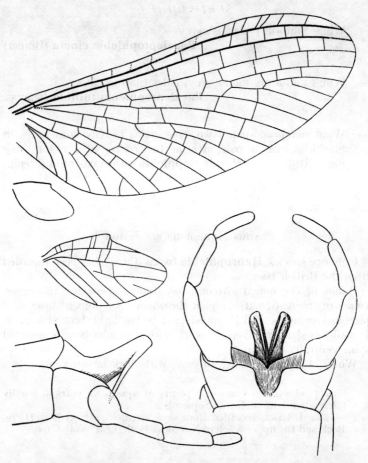

Fig. 24. *Habrophlebia fusca*, male. Wings (× 15) and hind wing enlarged; genitalia (ventral and lateral).

Fam. EPHEMERELLIDAE

Only one genus, *Ephemerella* Walsh, with two species, occurs in the British Isles.

Genus EPHEMERELLA Walsh

IMAGINES

1 Abdominal sternites I-VII or VIII each with black markings as in fig. 25N; general colour of male dark yellowish brown, of female yellowish. Penis-lobes slender, with a deep V-shaped excision between them (fig. 25N)— **Ephemerella notata** Etn.

Local, records mostly northern; moderately fast rivers, joining *E. ignita* in the lower reaches. Late May and early June.
Not recorded from the Lake District, but occurs in the River Eden in the neighbourhood of Salkeld.

— Sternites without such markings; general colour reddish brown, especially in female. Penis-lobes stouter, with a small U-shaped excision between them (fig. 25I)— **Ephemerella ignita** (Poda)

Common in fast streams and rivers, lowland and in the hills, either acid or alkaline in nature. April to September.
Lake District, common in rivers and small, stony streams up to at least 1,500 ft.

SUBIMAGINES

1 Wings pale greyish, with a yellowish tinge derived from the longitudinal veins, body yellowish, abdominal sternites marked as in imago— **Ephemerella notata** Etn.

— Wings dark greyish or blueish black, with blackish venation, hind wings sometimes paler, body olive brown in male, apple-green in female— **Ephemerella ignita** (Poda)

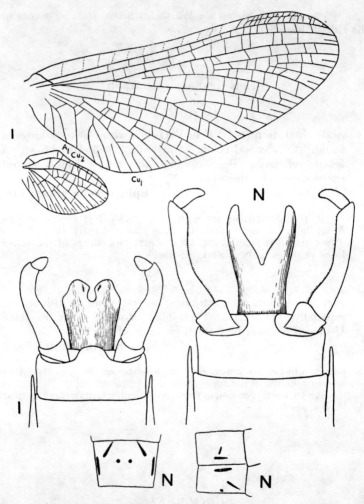

Fig. 25. *Ephemerella*. Male wings (× 9), genitalia (ventral) and abdominal pattern (ventral and lateral). I, *E. ignita*; N, *E. notata*.

Fam. POTAMANTHIDAE

Genus POTAMANTHUS Pictet

Potamanthus luteus (Linnaeus) is the only species of this family recorded from the British Isles (fig. 26). The general colour of the imago is yellowish, with a broad longitudinal dorsal brownish band. Abdominal tergites II-IX also with a black dot on each side at the base. Setae yellowish with the joints finely annulate with blackish brown. Legs yellowish with brownish markings. Wings hyaline, faintly yellowish, costal area of fore wing brighter yellow.

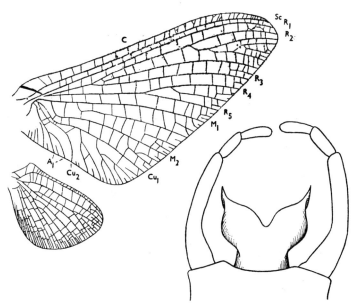

Fig. 26. *Potamanthus luteus.* Male wings (after Eaton) × 5·5 and genitalia (ventral).

The subimago has the wings yellowish, in the female tinged with greenish grey. Cross-veins at first pale, later becoming blackish. Abdomen yellowish, with a broad dorsal stripe and other markings as in the imago.

> Even less frequent in collections than *Ephemera lineata,* probably owing to its nocturnal habits. The subimago emerges during the

night and Eaton records that the imagines come to light. The nymphs occur in large, moderately fast rivers. R. Usk May 1955, R. Wye 1958 and older records from Laleham and Weybridge, River Thames, in July.
Not recorded from the Lake District.

Fam. EPHEMERIDAE

Only one genus of this family, *Ephemera* Linnaeus, occurs in the British Isles. It contains three species.

Genus EPHEMERA Linnaeus

IMAGINES

1 Abdomen ivory-white or light grey above, only the apical tergites commonly with brownish or blackish markings; tergites I-V either entirely without markings (particularly in female) or with cuneiform spot on each side, those on tergites I-II sometimes expanded to large quadrangular spots; on tergites VI-IX similar larger cuneiform spots, with between them often a pair of divergent lines. Wing markings dark olive-brown (figs. 1, 27, 28D)—
Ephemera danica Müll.

Common or abundant in lakes and rather faster-flowing rivers and streams, usually inhabiting waters with a lower average temperature than the preceding species, with a preference for alkaline waters. Mid-April to mid-September.
Lake District, lakes such as Windermere, Derwentwater and Esthwaite Water, in areas where there is a sandy or silted bottom, and in rivers and slow streams.

— Abdomen above yellowish, ochreous, or reddish brown, markings not as above— 2

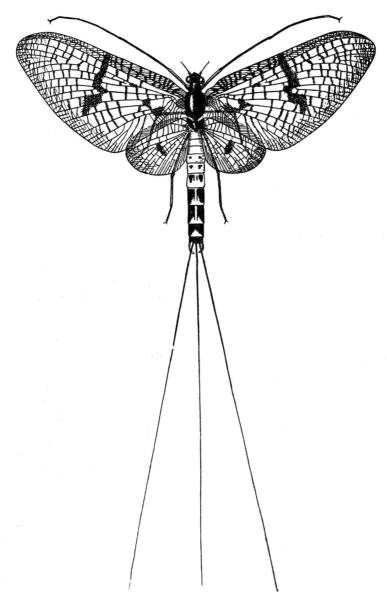

Fig. 27. *Ephemera danica*. Male imago × 3·25.

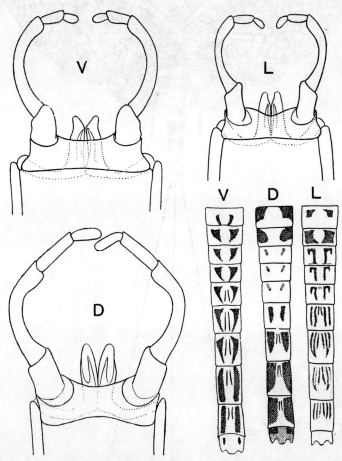

Fig. 28. *Ephemera*. Male genitalia and pattern of abdominal tergites. V, *E. vulgata*; D, *E. danica*; L, *E. lineata*.

2 Abdominal tergites V or VI-IX with three longitudinal black lines on each side. Wings pale yellowish brown, with small warm brown spots (fig. 28L)— **Ephemera lineata** Etn.

Scarce in collections. R. Wye in 1968 and some old records from localities on the River Thames, near Reading, and at Laleham, Weybridge and Teddington. July.
Not recorded from the Lake District.

— Tergites V or VI-IX with a broad somewhat triangular dark brownish band on each side, enclosing a pair of narrow longitudinal blackish lines. Wings suffused with pale warm brown, spots strong, reddish brown (fig. 28V)—
Ephemera vulgata L.

Moderately common in sluggish rivers with rather muddy bottom. May to mid-August.
Not recorded from the Lake District.

SUBIMAGINES

The subimagines may be separated by the abdominal markings as in the imagines. The wings are generally greenish grey, with blackish markings. In dried specimens this greenish colour fades, leaving the wings greyish in *E. danica* and somewhat brownish grey in *E. vulgata*.

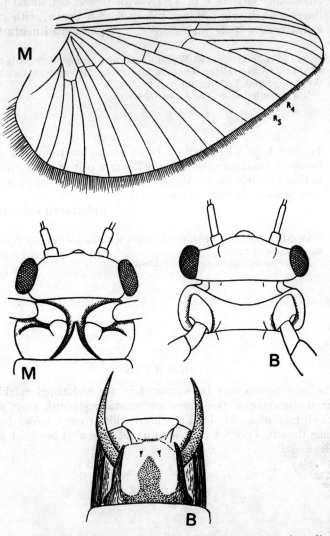

Fig. 29. *Caenis* and *Brachycercus*. M, *C. macrura*, wing ($\times$ 18) and ventral view of head and prothorax; B, *B. harrisella*, ventral view of head, prothorax and male genitalia. (Head and prothorax of each and male genitalia after Schoenemund).

Family CAENIDAE

1 Inter-coxal process of prosternum very narrow, two or three times longer than broad, so that the fore coxae are close together; second antennal segment not elongate (fig. 29M)—
CAENIS Stephens

— Inter-coxal process of prosternum very broad, twice as broad as long, fore coxae wide apart; second antennal segment three times as long as the first (fig. 29B)— BRACHYCERCUS Curtis

Genus BRACHYCERCUS Curtis (*Eurycaenis* Bengtsson)

Only one species, **B. harrisella** Curt., occurs in the British Isles, and should be readily recognised by the characters given in the generic key. The male genitalia are shown in fig. 29B.

The wings of the subimago are tinted with blackish grey, venation darker. Abdominal segments III-VIII have lateral filaments, which are long in the subimago and the male imago, shorter in the female imago.

Muddy parts of large rivers.
Not recorded from the Lake District.

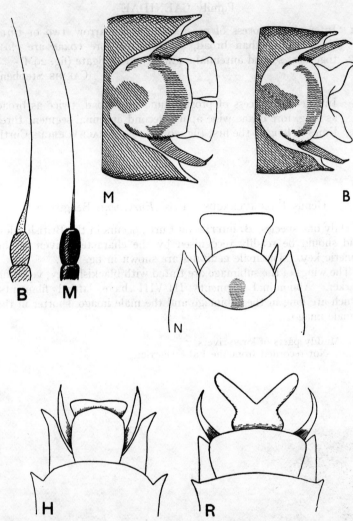

Fig. 30. *Caenis*. Male genitalia (ventral) and antennae. M, *C. macrura*; B, *C. moesta*; N, *C. robusta*; H, *C. horaria*; R, *C. rivulorum*.

Genus CAENIS Stephens

IMAGINES

1   Setae whitish, or yellowish white. A small, slender, finger-like, median process on apical margin of second abdominal tergite— 2

—   Setae sepia-grey or greyish white, sometimes annulated with blackish grey. No such finger-like process on second abdominal tergite— 4

2   All abdominal tergites marked or shaded with greyish; male forceps short and stout, about half as long as penis-lobes; forceps-base broad oval (fig. 30N)— **Caenis robusta** Etn.

Bred in 1954 from nymphs from Wheatfen Broad, a small and shallow broad, hardly more than a backwater on the R. Yare near Surlingham. The nymphs were found amongst *Ceratophyllum* and water-lilies. In the same year a few nymphs were found in Scoulton Mere, a large sandy-bottomed pond which has a good deal of vegetation, about 16 miles from Norwich on the road to Watton. It had been cleared of mud by a bulldozer four years earlier, and the presence of *Limnaea stagnalis* and *Planorbis cornuta* indicated that the water was fairly calcareous. After publication of a description of the nymph it was found in six other counties, generally in organic mud. In Holland *C. robusta* has been recorded from slowly flowing water (R. Ijssel near Gouda) in July.
Not recorded from the Lake District.

—   Usually not more than six basal tergites with greyish markings; male forceps longer and more slender, nearly as long as or longer than penis-lobes; forceps-base narrow oval or quadrate— 3

3   The first five or six tergites with blackish grey markings; thorax of male dark brown-black, of female lighter. Penis-lobes truncate, not divided or excised (fig. 30H)—
Caenis horaria (L.)

Common in large rivers, canals and lakes, where there is a silted bottom. June to September.
Lake District, I have seen examples from Windermere, Esthwaite Water and Three Dubs Tarn, and it no doubt occurs elsewhere. The duns emerge in the late evening, 7-9.30 p.m., G.M.T., in mid-June.

— Only the first three tergites with greyish markings, often faint; thorax light brown. Penis-lobes with a wide V-shaped excision* (fig. 30R)— **Caenis rivulorum** Etn.

More localised than the preceding species, inhabiting small, stony streams. June, July and September.
Lake District, Rydal Beck, Scandale Beck, Blelham Beck, River Kent near Staveley.

4 (1) Base of terminal antennal bristle conically dilated. Thorax black. A pyriform pigmented area on forceps-base, penis-lobes with a broad, sinuous excision at apex (fig. 30B)—
Caenis moesta Bengtss.

Occurs in lakes, rivers and streams, both alkaline and acid. The nymph prefers a sandy bottom with less vegetable detritus than does *C. horaria*. June to September.
Lake District, common in shallow, sandy bays of Windermere (and probably elsewhere in the Lake District), where I have seen large swarms flying in the early morning sunshine, 5.45-6.15 a.m., G.M.T., at the end of June. Some swarms extended from water-level to a height of about twenty feet.

— Base of terminal antennal bristle not dilated. Thorax black. An ovate piceous area on forceps-base, penis-lobes truncate†. (fig. 30M)— **Caenis macrura** Steph.

Common in large rivers and slower sections of smaller rivers, possibly with a preference for those of an alkaline nature. The imago flies in the early morning. June to August.
Not recorded in the Lake District.

SUBIMAGINES

The male subimagines of *Caenis* differ from the imagines chiefly in the much shorter setae, and should therefore be separable by the key above.

---

* The excision shown in Fig. 30R is exaggerated.
† The median slit in the penis lobes shown in Fig. 30M is probably accidental.

INDEX TO FAMI[LIES]
IN T[HE]

Ameletus Etn.
Arthroplea Bengtss.
BAETIDAE
Baetis Leach
Brachycercus Curtis
CAENIDAE
Caenis Steph.
Centroptilum Etn.
Cloeon Leach
Ecdyonurus Etn.
Ephemera L.
Ephemerella Walsh
EPHEMERELLIDAE
EPHEMERIDAE
(*Eurycaenis* Bengtss.)
Habrophlebia Etn.
(*Haplogenia* Blair)
Heptagenia Walsh
HEPTAGENIIDAE
Leptophlebia Westwood
LEPTOPHLEBIIDAE
Paraleptophlebia Lest.
POTAMANTHIDAE
Potamanthus Pict.
Procloeon Bengtss.
Rhithrogena Etn.
SIPHLONURIDAE
Siphlonurus Etn.

REFERENCES

Bengtsson, S. (1917). *Ent. Tidskr.* **38**, 184.

Blair, K. G. (1929). Two new British Mayflies (Ephemeroptera). *Entomologist's Mon. Mag.* **65**, 253-5.

Blair, K. G. (1930). *Ecdyonurus longicauda* Steph. (Ephemeridae) re-instated in the British List. *Entomologist's Mon. Mag.* **66**, 56.

Blair, K. G. (1930). A list of the British Ephemeroptera. *Entomologist,* **63**, 82.

Eaton, A. E. (1883-1888). A Revisional Monograph of recent Ephemeridae. *Trans. Linn. Soc. Lond.* (2) Zool. **3**.

Eaton, A. E. (1888). A concise generical Synopsis, with an annotated list, of the species of British Ephemeridae. *Entomologist's Mon. Mag.* **25**, 9-12, 29-33.

Edmunds, G. F., Allen, R. K. & Peters, W. L. (1963). An annotated key to the nymphs of the families and subfamilies of mayflies (Ephemeroptera). *Univ. Utah biol. Ser.* **13**, 1, 1-49.

Harris, J. R. (1952). *An Angler's Entomology,* London.

Hincks, W. D. & Dibb, J. R. (1935). Preliminary note on a probable addition to the British List of Ephemeroptera. *J. Soc. Br. Ent.* **1**, 75-6.

Kimmins, D. E. (1939). An Addition to the list of British Ephemeroptera. *J. Soc. Br. Ent.* **2**, 8-11.

Kimmins, D. E. (1942). Notes on Ephemeroptera. *Entomologist,* **75**, 121-5.

Kimmins, D. E. (1942). The British Species of the Genus *Ecdyonurus. Ann. Mag. nat. Hist.* (11), **9**, 486-507.

Kimmins, D. E. (1943). A species of *Caenis* (Ephemeroptera) new to Britain, with notes on the nymphs of some other species. *Entomologist,* **76**, 123-5.

Kimmins, D. E. (1954). *Caenis robusta* Eaton, a species of Ephemeroptera new to Britain. *Entomologist's mon. Mag.* **89**, 117-18.

Kimmins, D. E. (19
(= *Procloeon ru*
Entomologist's Gaz

Klapálek, F. (1909)
Deutschlands 8, 1-

Kloet, G. S. & Hinck
2nd edition (com

Lestage, J.-A. (1921).
des Insectes d'Eur

Mosely, M. E. (1921)

Mosely, M. E. (193
meroptera. *Ento*

Mosely, M. E. (193
Esb.-Peters not *E*
9, 91-6.

Müller-Liebenau, I.
Gattung *Baetis* L
Abwäss. 48/49, 1-

Needham, J. G., Tr
of Mayflies, with
Ithaca, New York

Rawlinson, R. (1939
Ecdyonurus venos
109, 377-450.

Sawyer, F. E. (194
appearance of *C[*
Trout Mag. **114**,

Schoenemund, E. (
Dahl, *Tierwelt Dt*

Ulmer, G. (1929).